EXPLORING the SCIENCE and BEAUTY of NATURE:

a photographic autobiography

by

RON S. NOLAN

PLANETROPOLIS PUBLISHING

222 Santa Cruz Avenue, No. 11
Aptos, California 95003
www.planetropolis.com

This book is an original publication of Planetropolis Publishing.

ISBN-13: 978-1-7379681-5-3

DEDICATION

Michael H.D. Dormer

Michael H.D. Dormer was a master of the creative process and the art of thinking outside the box. My novels and photo books are devoted to his memory.

****

Special thanks to
Julie Kay Adams for her
love and supportl

Table of Contents

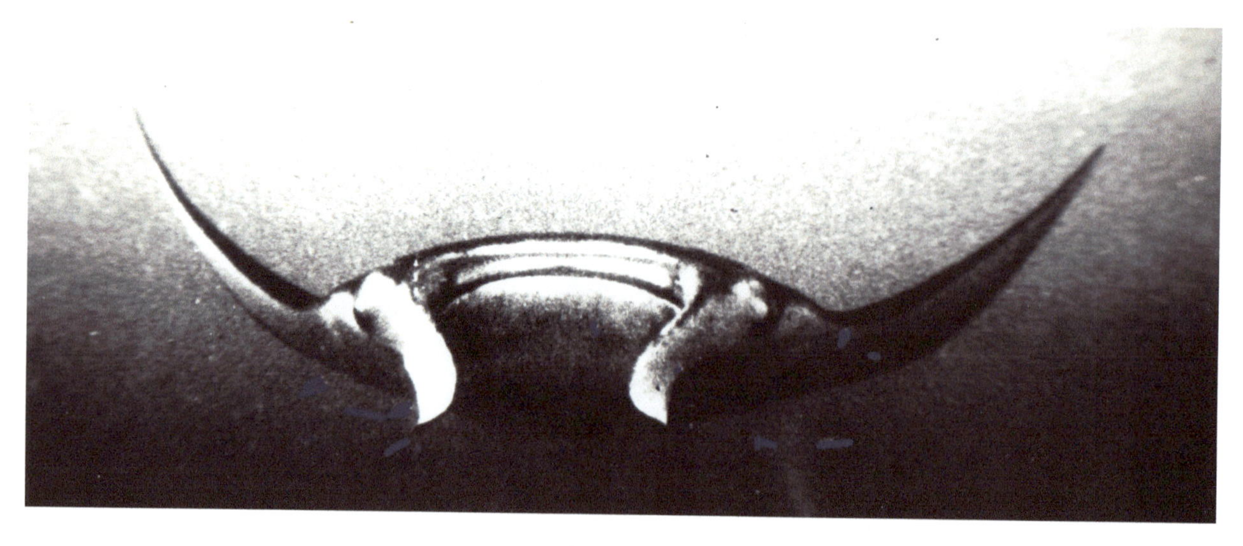

Manta Ray

This book is an autobiographical, photographic work of non-fiction that explores some of the amazing and wonderful experiences that I have had as marine biologist, entrepreneur, and photographer.

Yosemite Falls and Ferns

A second edition is planned that will take place in times and places before or after those described in this first book–including stories about some of the world's most famous celebrities and intellectuals.

CHAPTER 1
Journey to the Philippine Sea

Babuyan Island, Philippine Sea

After considering multiple options on how to proceed with an autobiographical book that employed the usual standard ancestral history beginning at birth and so on, I decided to save that info for later and start with some unique adventures that were fun, dangerous, and very unique—so after a segment about collecting manganese nodules from the Challenger Deep which is the deepest point of the ocean, along with rare specimens of bioluminescent deep sea fishes on the RV Melville in the Philippine Sea, I will turn my attention to a journey to one of the most radioactive spots in the world where I did my doctoral research.

So our first destination in this narrative will be in the Philippine Sea aboard the RV Melville then we will shift our story to Micronesia—specifically Enewetak (also often spelled Eniwetok) Atoll in the Marshall Islands where I studied reef fish ecology as a graduate student at Scripps Institution of Oceanography, University of California, San Diego.

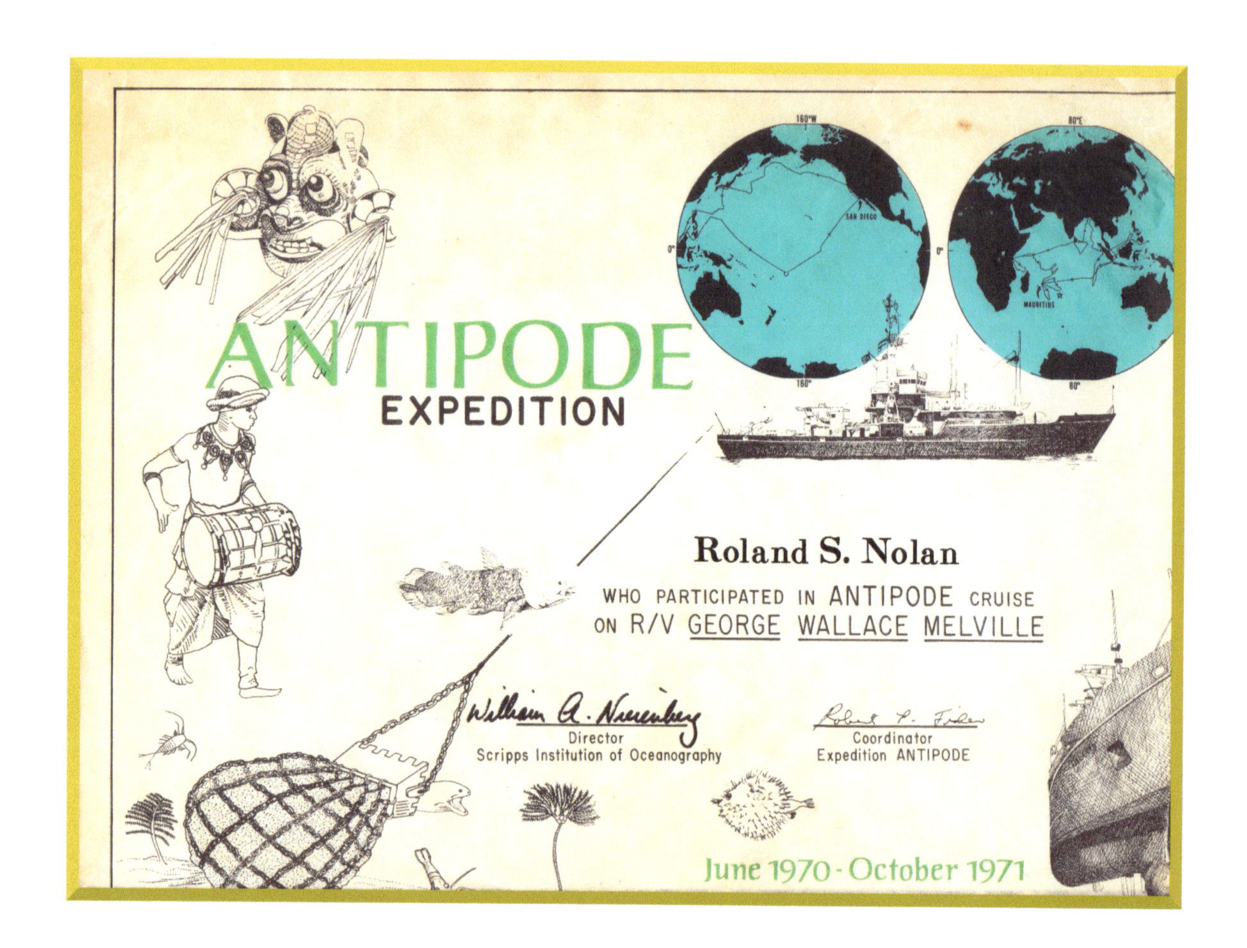

Research vessel Melville was built in 1969 with funding by the U.S. Navy, and subsequently owned by the Office of Naval Research and operated by Scripps Oceanography as part of the University-National Oceanographic Laboratory System (UNOLS). Configured as a general-purpose oceanographic research vessel, Melville supported a wide range of scientific activities across every discipline of oceanography, and involving capabilities as diverse as deep-towing cameras, deploying massive moorings, precisely maneuvering remotely-operated vehicles thousands of meters below the ship, and launching (and recovering!) unmanned aircraft used to measure gravity and atmospheric physical properties.
Source: EMERITUS: R/V MELVILLE

Author Breaks the Rules

Roll, rubble, tumble, jumble
freedom to be free......——.......
Brind, gessusch, gading, gagi
structure?....That's not for me.

To do as expected is not to be free...
shananzel, garbunzel, quanastal, kabee.

To do as expected is not for me.
Mabu, Kuba, Zaining, Smozee.

'Tis the call of the...
lecherous life apprenticee!

*(Photo taken of me diving off the
RV Melville on a very calm day in the Philippine Sea.)*

CHAPTER 2
The Science & Beauty of Bioluminescence

There is an entire world in the darkness of the deep sea in which bioluminescence is employed as a lure to attract prey and mates.

With the assistance of the crew members and the lab techs who provided assistance in collecting specimens, we had great success in observing and recording these amazing bioluminescent adaptations.

Gonostoma elongatum

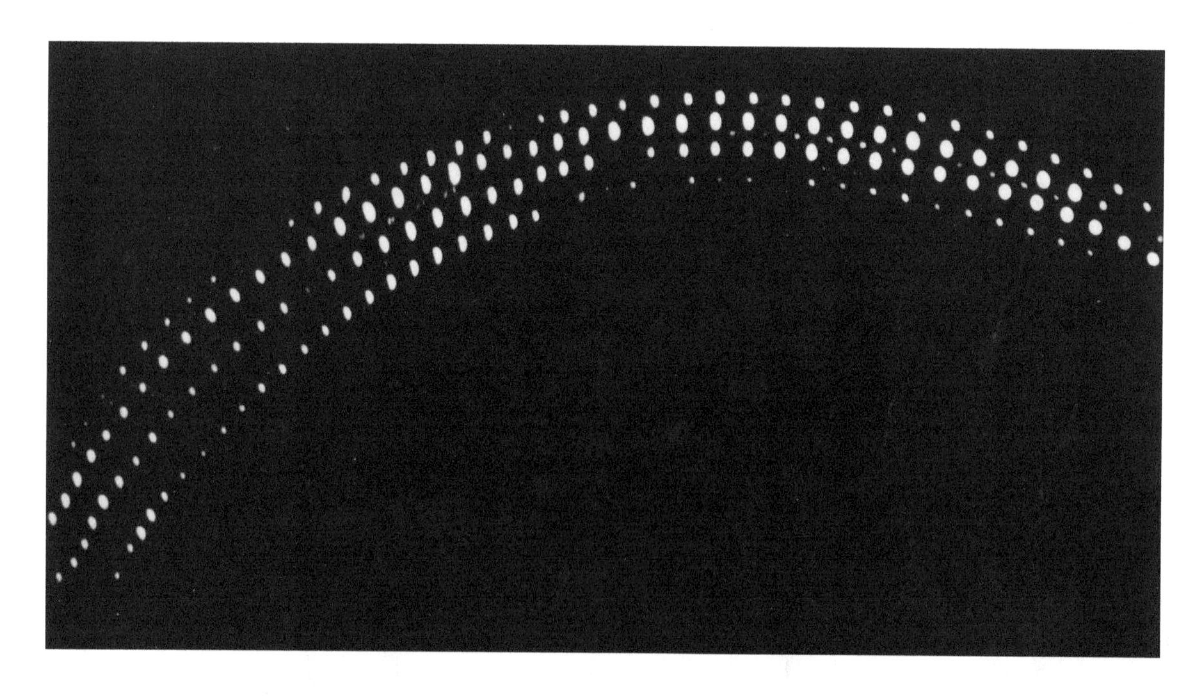

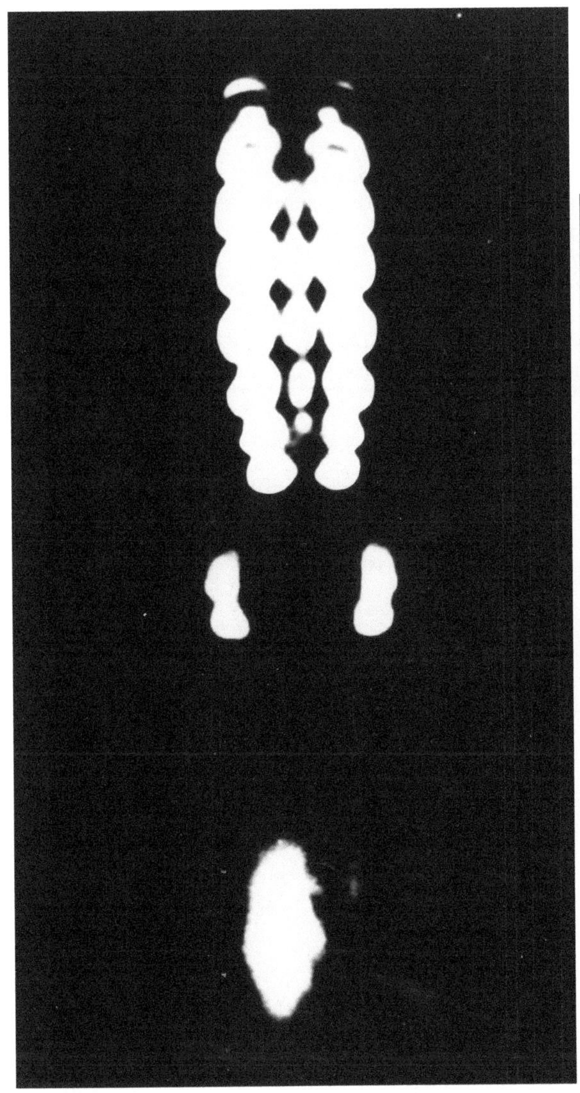

Sternoptyx diaphana

After the Antipodes Expedition, Professor Rosenblatt and I discovered and described a new species of anglerfish which we named *Lasiognathus waltoni* after Isaaz Walton's work the *Compleat Angler*. There was only a single specimen available for our review of the genus. The males are much smaller, parasitic and attached to the female. They serve as a source of sperm for reproduction and depend upon the female for nutrition and safety.

A Review of the Deep-sea Angler Fish Genus *Lasiognathus* (Pisces: Thaumatichthyidae)

Ron S. Nolan and Richard H. Rosenblatt

The first specimen of *Lasiognathus saccostoma* (and the genus) from the Pacific is reported. The esca of this specimen closely resembles that of the holotype and also that of *L. ancistrophorus* Maul (here placed in synonymy with *L. saccostoma*). A new species of *Lasiognathus* from the North Central Pacific is described, based on its unique angling apparatus. The original description of *L. beebei* Regan and Trewavas (a painting) is reconsidered in light of a more recent specimen of *Lasiognathus* sp. collected off Madeira. The escal configuration and body proportions of *L.* sp. agree closely with that of *L. beebei*. Feeding habits are briefly discussed.

THE ceratioid angler fish genus *Lasiognathus*, previously known from five specimens, consists of three nominal species, all from the Atlantic. The collection of a female *L. saccostoma* Regan during the Piquero Expedition of Scripps Institution of Oceanography to the eastern tropical Pacific constitutes the first record of the genus from the Pacific.

A second specimen of *Lasiognathus* was captured by Scripps' Cato Expedition in the North Central Pacific. This specimen differs sufficiently in escal structure to warrant specific separation. Five additional specimens from the Pacific have also been taken, unfortunately all lacking the angling apparatus.

Materials and Methods

The specimens on which this study is based are housed in the Vertebrate Collection, Scripps Institution of Oceanography (SIO). Esca length

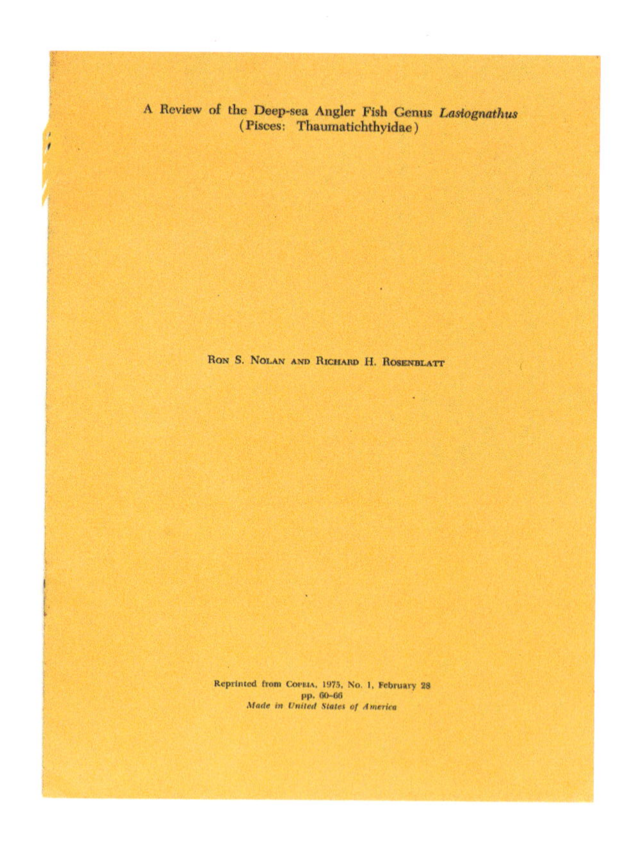
A Review of the Deep-sea Angler Fish Genus *Lasiognathus*
(Pisces: Thaumatichthyidae)

Ron S. Nolan and Richard H. Rosenblatt

Reprinted from Copeia, 1975, No. 1, February 28
pp. 60–66
Made in United States of America

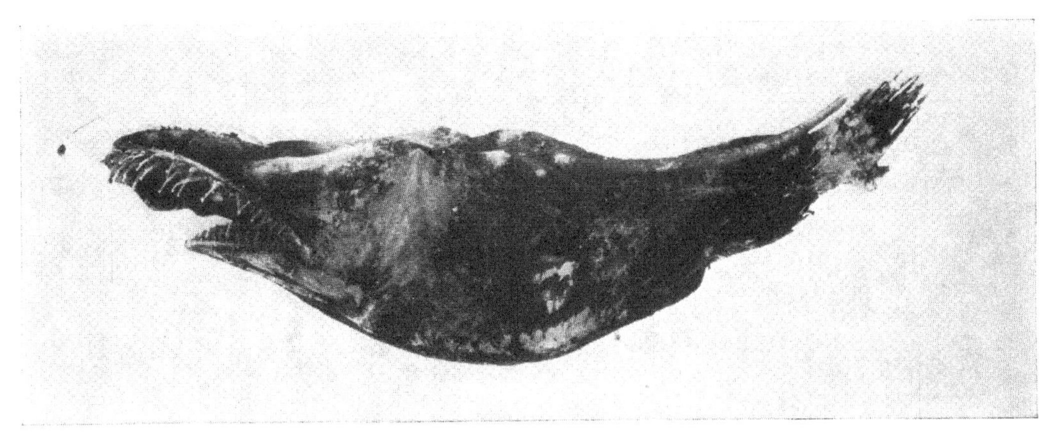

Fig. 1. Female *Lasiognathus saccostoma* Regan (95.5 mm SL, SIO 69-342).

was measured from the distal tip of the esca (the tip of the hook-bearing filament to the tip of the terminal filament, if present) to the proximal base of the escal bulb (including bulb). Tooth counts include small replacement teeth and empty sockets. Collection data for the specimens examined are as follows: *L. saccostoma*, SIO 69-342, 17°42.0′S–110°20.0′W, 10′IKMT, 0–1100 m, 3/29/69; *L. waltoni*, holotype, SIO 72-373, 30°39.1′N–155°23.4′W, 0–1350 m, 6/24/72; *L.* sp., SIO 73-305 (Teuthis), 21°16.0′N–158°19.6′W, 625–725 m, 2m Tucker Trawl, 6/29/72; *L.* sp., SIO 73-306 (Teuthis), 21°19.9′N–158°15.6′W, 775–950 m, 2 m Tucker Trawl, 2/20/71; *L.* sp., SIO 73-307 (Teuthis), 21°22.6′N–158°48.0′W, 800–1000 m, 2 m Tucker Trawl, 8/2/72; *L.* sp., SIO 70-343, 18°6.2′N–119°7.9′E, 0–1850 m, 2 m IKMT 9/17/70; *L.* sp., SIO 73-159, 31°1.9′N–155°4.4′W, 0–1450 m, 2 m IKMT 2/10/73.

Taxonomy

The genus *Lasiognathus* may be characterized as follows: body elongate, relatively shallow; no visible dermal spines; jaws greatly produced, upper jaw overhanging and extending markedly beyond lower jaw, supplied with elongate, narrow, down-curved acicular teeth; sphenotic, quadrate and angular spines well-developed; basal bone and illicium elongate, terminating

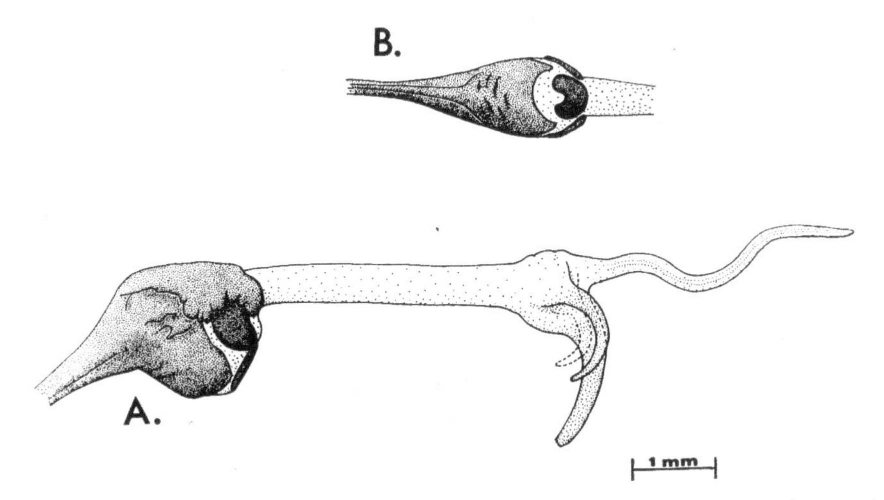

Fig. 2. Esca of *L. saccostoma* (SIO 69-342). A. Lateral view. Dorsal flap wrapped around upper half of bulb (pore and pigment blotch beneath dorsal flap); right lateral pigment blotch just below margin of flap immediately below lateral blotch is ventral blotch (seen edgewise). B. Escal bulb as seen from below. Ventral pigment blotch lies between lateral pigment blotches.

CHAPTER 3
Tributes to Richard H. Rosenblatt, Robert K. Johnson & Carl L. Hubbs

Dr. Richard Rosenblatt (my major professor) was Chief Scientist of the expedition and did a great job keeping our spirits up when we ran smack dab into not only one typhoon, but the same one twice. That's why Bob Johnson and I entitled the presentation that we later made at Scripps as "Over the Rail in the South China Sea."

In one of Dr. Rosenblatt's and my last e-mail message exchanges in which I expressed my concern that the argument presented by the *Intelligent Design* supporters regarding how difficult it would have been for the evolution of vision unless all the components (nerves, lenses, retina, sensors, etc.) were available simultaneously.

However, Dick did not appear to be concerned about this 'chaos' and as I recall merely said, "I believe in science. It will find the answer." That was my last interaction with the great man. One of the editor's at Copeia asked me to provide a comment about Dick's contribution to marine science. I immediately recalled that Dick always seemed to be most comfortable sitting in the Scripps Fish Collection seated among the thousands of jars packed with fish specimens and that each jar, small or large, had a story which he loved to tell to his students, techs and visitors.

Both of us being born and growing up in the Mid-West made us very appreciative of our marine biology careers. Not a day went by in which we didn't gaze at the beauty and energy of the ocean just steps away. I was really happy and honored that Dick asked me to come along on the month long Osaka to Manila leg of the Antipode Expedition. After we landed at Tokyo International, we took a commuter flight to Osaka where the RV Melville had just arrived and the crew ready for a bit of a holiday and relaxation–which we were eager to join in on and hear their stories.

I was very pleased to check out the well equipped darkroom where we planned to conduct our bioluminescence study on fishes that we would capture literally miles below the surface. I was also fortunate to have the funds to purchase my very own Pentax 35 mm camera upon which Dick taught me the basics. My very first shutter click was the gorgeous red sushi bar sign below.

Dick took me and my fellow student Bob Johnson on a commuter train ride to the Deer Park and the Toudaiji temple in Nara where we visited the immense Buddha statue.

The Sika deer are regarded as messengers of the gods and seemed to take an interest in Dick.

Bob Johnson seated next to Dick, was a brilliant scientist and a good friend of mine. Marine science severed a tremendous loss when Bob passed at an early stage in his career.

This is the very first picture that I have ever taken with a 35 mm camera. Dick was there and pointed out the workings of the timer and light meter.

CHAPTER 4
Journey to the Earth's Mantle

Now I will address my research in the Marshall Islands in Micronesia which all started when Scripps Professor John Isaacs a brilliant outside-the-box innovator and entrepreneur, wanted to reconnect one of the drill holes that research engineers aimed at reaching the Earth's mantle, often called the "Project Mohole."

The idea was that the mantle was much closer to the surface beneath islands than continents and many previous attempts to reach the mantel, including the one on Runit Island on Enewetak Atoll had been attempted in several locations, but none had yet been successful.

Enewetak Atoll, Marshall Islands, Micronesia

IR Photo of Cactus and La Crosse Nuclear Test Craters on Runit Island

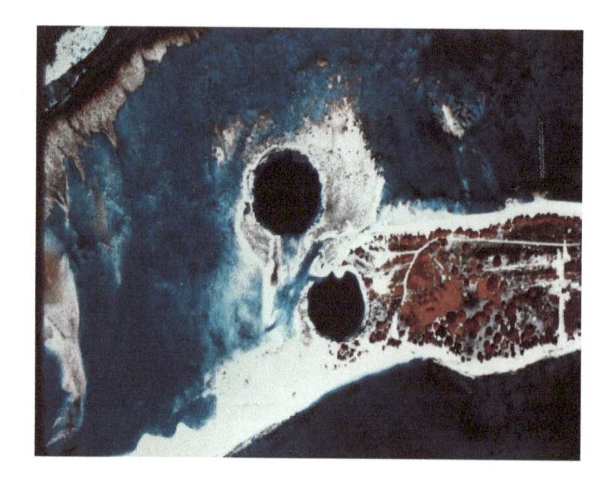

Professor Isaacs offered me the awesome opportunity to survey the fish populations of two nuclear test craters on the same Runit Island with the goal of demonstrating that there had been no long term impacts on the reef fishes that were likely inhabiting the craters.

As he put it, "Put the trust back in the Trust Territory."

It was a fun and challenging venture as I had to learn how to charge my SCUBA tanks with a portable generator along with no restrooms or plumbing. Luckily, Charlie Stearns pitched in as the chef, otherwise we would have had some serious, dietary issues.

In order to provide safety from radiation exposure. the Atomic Energy Commission sent an agent who camped out in a tent while the rest of us slept on cots in the Quonset Hut that had housed the teams in charge of monitoring nuclear explosions. We were all ordered to wear radiation badges and the AEC rep had us stop by his tent daily so he could scan our feet with a Geiger counter!

I think the most bizarre aspect was the rat population. They were all over the island and fearless of humans, running in front and along side our truck in broad daylight. We were probably the first humans they had seen and the food we brought with us was a major attraction.

One night they attacked us by climbing through one the windows. As they began scampering beneath our cots, one of the techs, Ron McConnaughey, started impaling the invaders with a pole spear which led to thrashing and screaming to the point that no one could ignore, so we started the generator and turned on the interior lights which ended the attack.

We enjoyed two weeks on Runit and spent several hours each day diving in the craters surveying reef fishes and mapping the bottom type with occasional visits by harmless, nurse and blacktip sharks.

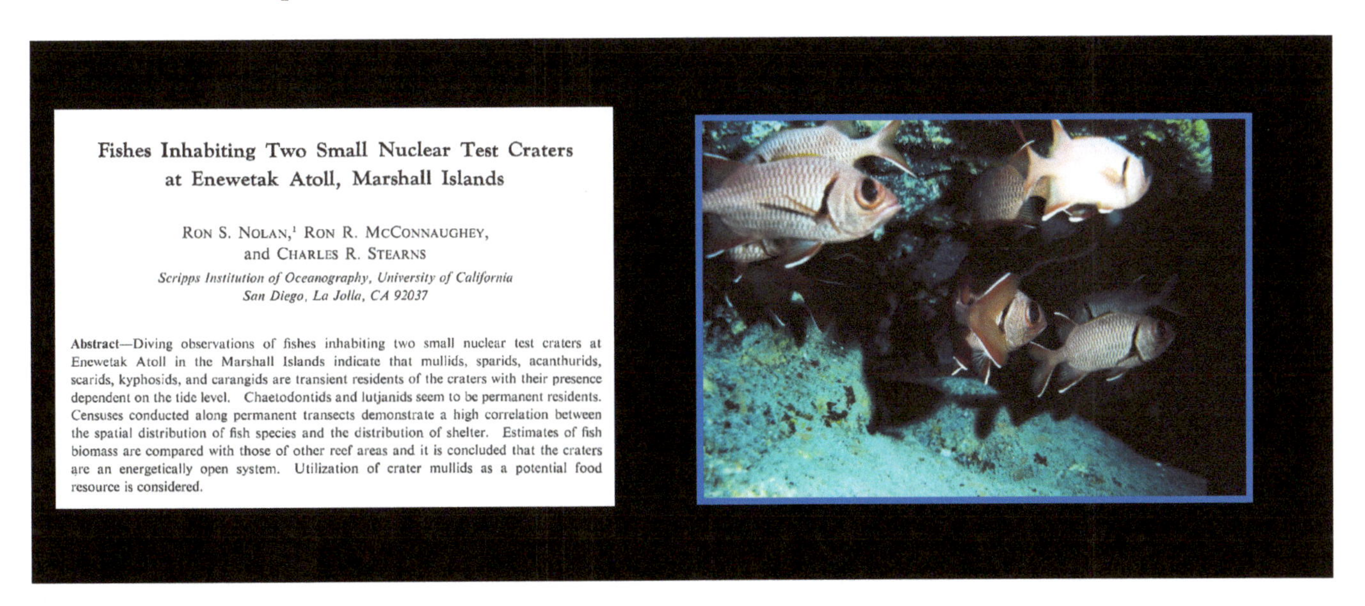

Fishes Inhabiting Two Small Nuclear Test Craters at Enewetak Atoll, Marshall Islands

RON S. NOLAN,[1] RON R. MCCONNAUGHEY, and CHARLES R. STEARNS

Scripps Institution of Oceanography, University of California San Diego, La Jolla, CA 92037

Abstract—Diving observations of fishes inhabiting two small nuclear test craters at Enewetak Atoll in the Marshall Islands indicate that mullids, sparids, acanthurids, scarids, kyphosids, and carangids are transient residents of the craters with their presence dependent on the tide level. Chaetodontids and lutjanids seem to be permanent residents. Censuses conducted along permanent transects demonstrate a high correlation between the spatial distribution of fish species and the distribution of shelter. Estimates of fish biomass are compared with those of other reef areas and it is concluded that the craters are an energetically open system. Utilization of crater mullids as a potential food resource is considered.

Measuring Coral Cover and Recording Fish Populations

CHAPTER 5
Creating Artificial Reefs

During several dives in the atoll lagoon, it was obvious that fish population levels were determined by the availability of certain species of acropora corals where they took shelter when predators approached and at night when the predator eels became most active. I wondered if I could learn more about this connection by building artificial reefs– even though I had no idea how this might become an actual research experiment,

I put together a proposed scope of work and funding budget for travel and equipment, submitted it to Professor Isaacs who agreed to cover the costs which were considerable. So over the next three years, I made thirteen month long trips with the help from lots of wonderful friends at the Enewetak Marine Biological Lab on the Main Island and the Air Force at Hickham Base in Honolulu who gave me a green light for travel on their Boeing C-141 cargo planes which made weekly flights to Kwajalein and then on to Enewetak.

Natural Patch Reef
Dozens of Species of Reef Fishes
Inhabit Patch Reefs

'Artificial Reef Factory' operated by my pals Tom Dana and Ron McConnaughey

Newly Installed Artificial Reef

Schools of Damselfish (Chromis species)
Colonized the Artificial Reefs

Banded Coral Shrimp

Tunicates, Algae and Sponges Colonized the Tubes and Became Food for Reef Fishes and Invertebrates.

Sewing a net to create a "Predator Exclusion Cage"

Cages on several of the artificial reefs were constructed to provide a refuge for reef fish larvae from predators. Significant changes in community structure were observed.

One of the Predators

Black Jack (Caranx lugubris)

I am intrigued by the widespread similarities of schooling behavior in diverse species ranging from small fish like those pictured above to pelicans and even antelope and crows and herds of elephants. The link to surviving attacks by carnivores and showing off their male attributes seems to be programmed into their genetic code as well as learning from their peers. Do certain individuals dominate the leadership, or do they all get a chance when the school flips direction and the straggler at the back of the flock suddenly becomes the leader? How often does this happen?

One Hundred Crows

Floating with the wind as it struck the cliff and roared upward, providing the flock of crows with a break from gravity and a free ride across the sky.

One hundred crows wheeled and swung, dove and climbed in unison as each of the young males waited impatiently for his chance for a few seconds in which he would lead the flock and the others would instinctively follow.

After that, he knew that he had to relinquish command as his rivals waited for their turn to totally command the flock.

Shortly after my last visit to my study reefs in 1975, I moved to Hawaii to continue my reef studies under the guidance of Dr. Leighton Taylor who was to become Director of the Waikiki Aquarium. In 1977, Cactus Crater was filled with 95,000 cubic yards of radioactive debris and covered with 18" of concrete. Today it is questionable if the dome is leaking radiation.

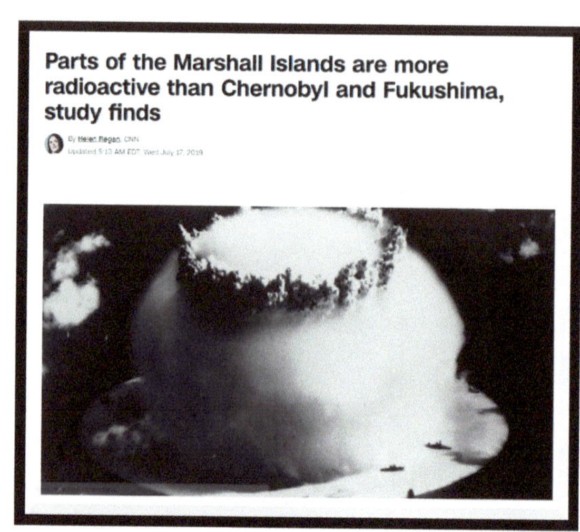

Parts of the Marshall Islands are more radioactive than Chernobyl and Fukushima, study finds

By Helen Regan, CNN

Dark Shark

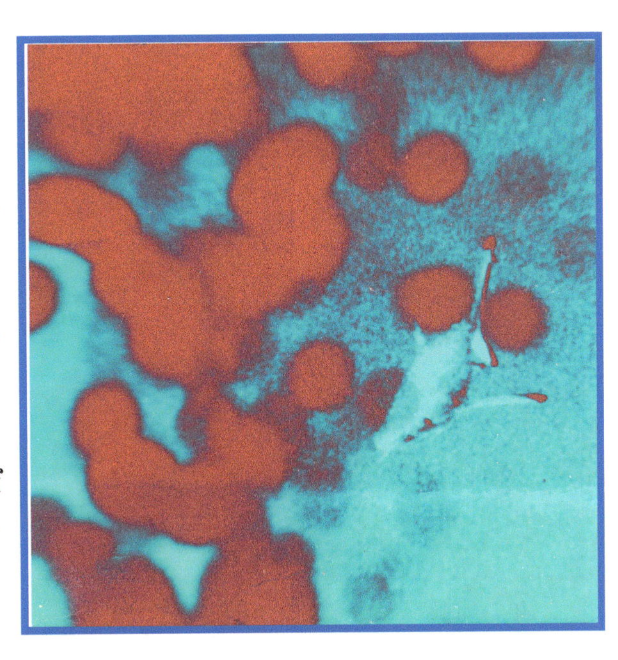

On Dark Shark—sensuous
omni-whore of the down there.

Come from the nothingness of lonely
fluid and slip nearer each orbit.

Closer—then to dynamite away,
then…
once again, The Orbit.

She tosses—flashes the beauty of
her steel flesh, perfect in form…
waiting to touch.

The Scorpio Lady of Death
——she Orbits.

Each time again and again, she the
lusty fox caressed by her universe
mother.

Circling closer, nearer and nearer
——tightening the Orbit!

Now I know that she my life must
decide,

and still closer the Orbit.

Wait… Where...?

Spin around, she is nowhere!
Gone?

No too late!

From below the Master comes
and takes...

Full Comes the Orbit!

Rhett McNair, former professional bull fighter and diving buddy developed a powerhead for defending divers from dangerous sharks. Here I photographed him testing his newly designed powerhead on a large, dangerous tiger shark that had been captured for the experiment.

CHAPTER 6
Journey to the Island of Hawaii

My Memory Lane, Looking across the Alenuihaha Channel toward my home on the Big Island.

If you glance at the image above, the photo looks out from Maui across the Alenuihaha Channel towards the North Shore of the Island of Hawaii. I made the exaggerated footprints by carefully walking forward and then backwards twice and then setting a camera on a tripod and using a timed exposure. I am the person sitting on the beach on Maui where the camera was located. At that time I was on a business mission and longing to get to my home in the distance. I have a large mural of this scene which I see every day in my computer lab where I am writing from right now and still feeling the longing to return to one of the most lovely places in the world. So now I will take you there. I will try to make it a whale of a good time.

Rural Kohala

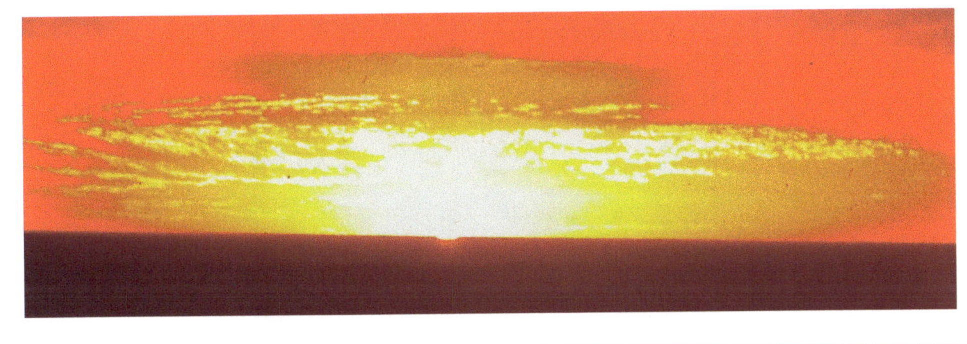

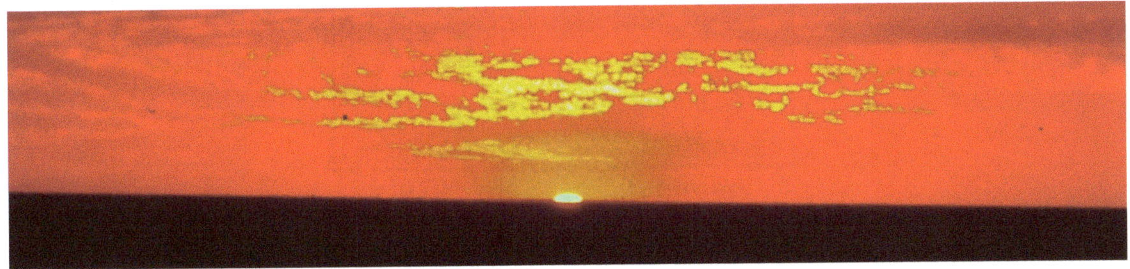

Green Flash (brings good luck!)

Waipio Valley, Late 1970's

Some places are so special that I don't want to reveal them.

CHAPTER 7
Journey to Parker Ranch

Parker Ranch, located on on the Big Island, is huge. (Big Island...Big Ranch) consisting of 130,000 acres of lush grasslands that thrive year round on two major peaks; Mauna Kea and the Kohala Mountains. The ranch's commercial beef operations support 26,000 Charolais (a French breed) and Angus cattle.

Tony Smart was a good friend. When I knew him, he and his family had assumed a key role in the Parker Ranch management. I don't remember exactly how it happened, but shortly after I rented a beach house near Puako, I was chosen to be the photographer for the girls rodeo team. They were highly skilled at calf roping and barrel racing and won the championship.

CHAPTER 8
Joy of Living on the Big Island

Paintings by the Gentle People of Waipio Valley

My personal and business life in Hawaii was exciting, highly involved, and with long lasting favorable impacts. The first two years were spent on the Big Island; in this old style plantation house near Puako where I established Ocean Research Consulting and Analysis (ORCA) my trade name for my consulting services as a marine biologist diver eventually with 3,000 hours of underwater fish surveys in Hawaii and Micronesia.

On the Island of Hawaii, also known as the *Big Island*, the plantation house that I rented a home was right next (ten yards from the back-yard gate to the sea—no kidding!) to a gorgeous beach that was located between the small village of Puako to the south and Kawaihae Harbor to the north within a short distance to the high end Mauna Kea Beach Hotel.

People that I hardly knew, contacted me to see if they could "stop by" and I usually had someone in the second bedroom despite my true feelings. Some were most welcome, like the artists that I remember as the "Gentle People from Waipio Valley" who worked together on the beautiful plantation house and the small lava flow island paintings which are still hanging on my walls as I write this autobiography.

Wailea Bay,
Beach 69 and Lava Island

While I lived next to Beach 69, the name locals used to refer to the unnamed, rough road only identified by the number 69 on the telephone pole that led to my house, I was evacuated once due to a tsunami warning and on another occasion I could see a major eruption on Mauna Loa, and since I was surrounded with fields of lava and there was even a lava island a few meters offshore from prior flows, I stayed up late to watch the spectacular fireworks! This summit eruption didn't cause any damage, but in later years their have been several catastrophic events at the Volcano National Park on the eastern side of the island.

Mauna Loa Summit Eruption
(July 5th 1975)

Lava Island
(located in front of my house)

My friend Mark Murray provided sailing trips to Mauna Kea Beach Hotel guests on his luxurious catamaran. John Denver was a regular patron and we spent many get-togethers exchanging stories about reef diving and flying sail planes, which we both loved to do. The highlight of my time with John was when he and his band brought their instruments onboard Murray's catamaran and as the sun set on a calm day, Mark dropped the sails. As we drifted, John and the band gave us a concert that we will never forget. I hope there is a heaven and John is there, smiling and playing his guitar.

CHAPTER 9
Aquarium Fish Collecting Impact Study

Dr. Leighton Taylor, Jr. Principal Investigator

In the late 1970's Dr. Leighton Taylor, shark expert and environmentalist, conceived of a grant proposal that was funded by Sea Grant which included a stipend for me. At the time he was working at the Hawaii Cooperative Fishery Research Unit of the US Fish and Wildlife Service and was able to provide a skiff for me to have access to the nearby reefs where I was to do fish surveys to investigate whether the increasing number of aquarium fish collectors were threatening reef fish populations.

Yellow Tang

It was apparent that the need to understand and protect reef fish populations in Hawaii was a sensitive environmental issue with conservation advocates and commercial fishermen at odds with the growing number of aquarium fish collectors. Commonly collected reef fish were colorful tangs, butterfly fishes and angelfishes. Photos of species focused on by collectors follow.

Blue Striped Butterflyfish

Two years after we began the project, we presented our findings at a meeting arranged by officials and the public in which Leighton and I made presentations of our survey data and stated our opinion that the aquarium fish collecting business should be regulated in a fashion similar to the commercial fishing industry and especially that marine preserves should be established where fishing of any sort, aquarium or commercial, would not be permitted. The response was mixed.

Leighton went on to become the Director of the Waikiki Aquarium and continued to find projects for me to work on, for example, saving a giant whale shark trapped in a lagoon and on the verge of starvation!

And...two and a half decades later, Leighton and I published an analysis of our underwater survey methods in the journal *Naturalista* with the goal of standardizing reef fish census methods for others to use in the future.

Naturalista sicil., 1999, Vol. XXIII (Suppl.),.pp. 261-283

RON S. NOLAN & LEIGHTON R. TAYLOR, JR.

AN EVALUATION OF TRANSECT METHODS FOR FISH CENSUS ON SHALLOW REEFS

ABSTRACT

The estimation of shallow water fish populations by SCUBA divers swimming along transect lines is a widespread method that has not been critically analyzed and for which standardization is needed. During a two-year field study of Hawaiian reef fishes, we assessed modifications of the method to determine the most efficient and accurate means of conducting transect censuses.

We concluded: (1) increasing line length increased the accuracy of the population estimate (both number of species and relative abundance) but an optimum of effort and accuracy was achieved with a 50 m. long transect; (2) at least two replicates per transect were necessary to estimate 75% of the species present; (3) no significant difference in population estimates of selected species was noted between transect runs which recorded all species present and those which recorded only selected species, or between runs on which data were recorded on a slate by hand and those on which an underwater tape recorder was used; (4) trained observers on "blind" trials were able to detect differences in the number of individuals of certain species before and after removal by spearing; (5) seasonal variations in population size were detected on the same transect lines after a one year interval.

The transect method of e<stimating fish populations is accurate if care is taken in selecting transect length, a standardized method suitable for the habitat is used, and sufficient numbers of replicates are conducted by trained observers.

CHAPTER 10
Coral Reef Survey of the West Hawaii Coast

In 1978 Dr. Dan Cheney, marine biologist at UH's campus in Hilo, and I were chosen by the US Army Corps of Engineers to survey the marine life of the entire West Coast of the Island of Hawaii. This was a very ambitious undertaking that traversed hundreds of underwater miles over a year long period and ranged from the northern harbor in Kohala to South Point —which is the southern most point of the U.S.

We produced a lengthy manuscript that described each study location and a record of fish and invertebrate populations as well as the type of bottom substrate. We also produced a West Hawaii Coral Reef Atlas that used satellite imagery to mark the exact location of each of our census locations.

South Point (Kalae)was the last of our survey sites and was one of the most unique in terms of marine life and geographic location–it is the southernmost point of not only the State of Hawaii, but also of the United States of America!

View looking toward North from South Point

View looking South toward South Point

Commercial Fishing at South Point

Since the sea bottom dramatically drops off right adjacent to the cliffs, fishermen are able to secure their boats with lines to the shore and only rarely need to start their outboard motors.

Reef Fishes and Invertebrates Observed in West Hawaii Survey

Photo by Robert S. Kiwala
Fish Collector & Submarine Captain
Scripps Aquarium

Jacks on Patrol

Nurse Shark

Two Spot Red Snapper
Lutjanus bohar

Hawkfish (*Paracirrhites arcatus*)

Longnosed Butterflyfish
Forcipiger flavissimus

Tube Worm

Crown of Thorns starfish are attacking reefs globally.

CHAPTER 11
High Altitude SCUBA Dive

Dave Norquist gazes at the snowstorm on the Big Island peaks while he prepares to dive off the Kona Coast.

I was an alpine ski enthusiast and therefore pleasantly surprised when I moved to the plantation house at Beach 69 on the Kohala Coast of the Big Island and discovered that winter storms would bring lots of snowfall to the top of Mauna Kea and there was even a tow lift and a ski club. Imagine skiing in the early morning then going down to beach and drinking cold beer to cool off in the hot afternoon sun!

Lake Waiau, 13,020 feet elevation, Mauna Kea, Island of Hawaii

On my first Hawaiian ski venture, I noticed in the distance a lake below the summit of Mauna Kea and learned that it was called *Lake Waiau.* In the next few days I did some research and discovered that Hawaiian folklore claimed that Lake Waiau was bottomless and led to the other side of the world. I spoke with several local residents and scientists at the University of Hawaii campuses in Hilo and Manoa and no one knew what really was at the bottom of the lake.

So I started planning a mission to find out more about the lake–the physics, health, and mental aspects, as well the unique and dangerous risks of extremely high altitude SCUBA diving and I pitched to the local television news station who agreed to provide live coverage. After that, I found a group of experienced divers that wished to be part of the crazy venture that I was organizing as well as an accomplished and enthusiastic researcher from UH. Who volunteered to assist in the collection and analysis of any samples that we might collect.

We had to keep in mind that we would have to go a considerable distance on foot because the road stopped at the observatory. It would be a challenge to trek through the snow and patches of frozen lava to get from our parked vans to the lake.

Before we began the hike to the lake, we put on the insulated dry suits that were loaned to us by a dive shop in California. Our usual wet suits would not offer nearly as much warmth and comfort as the specialized dry suits.

Our priority was to get to the lake as quickly and effortlessly as possible taking occasional inhalations of pure oxygen when we began to feel breathless or dizzy. Keep in mind that Part 91 of General Aviation Rules "requires flight crews to use supplemental oxygen for any part of a flight that exceeds thirty minutes above a cabin pressure altitude of 12,500 feet mean sea level. Diving in Lake Waiau would surely test our willpower as well as our circulatory systems.

One of the problems we faced was that the road at the top of the volcano didn't access the lake, so we had to disembark and drag our heavy SCUBA tanks and weight belts on sleds–which was a considerable exertion. Luckily my buddy Mark Murray rounded up some oxygen bottles and masks.

Mission Accomplished!

April 4th, 1976

The following is the sequence of actions that I personally employed.

I carefully waded into the water, put on my mask, inserted my mouth piece, and then submerged.

The first breath I took seemed normal, but my respiratory system had just plunged from 13,020 feet to a minus zero feet (sea level pressure) in just one breath.

That was not a serious problem, but after a thorough exploration, I opted to surface to check my condition. I understood that the gasses dissolved in my system had been under a relative high pressure equivalent to sea level (more when I descended deeper) and that I nearly instantly went from sea level on the reef at Kona to Lake Waiau at 13,200 feet in a couple of seconds–or less when I started breathing ambient gases. I did feel weak and had a sense of vertigo–both of which went away as we descended down the mountain road towards Waimea and Kona.

With as little exertion as possible after accomplishing our mission goals, we all needed to return back down to sea level ASAP or we could contract a severe–even life threatening–case of decompression sickness...aka *the bends* in which our blood could actually begin to boil. Therefore, racing back down to sea level as fast as possible was the highest priority.

Although we did set a high altitude SCUBA dive record in 1976 in Lake Waiau at an altitude of 13,200 feet, the highest dive currently is 6,395 m (20,980 ft 11 inches) achieved by Marcel Korkus (Poland) on 13 December 2019 at Ojos del Salado, Argentina.

Now you may wonder what we discovered. Sorry, I agreed not to tell. End of story, but we all returned safely and Diver magazine published the article that I wrote and the news team sent us the following confirmation:

KHON-TV INCORPORATED

1170 AUAHI STREET
HONOLULU, HAWAII 96814
TELEPHONE (808) 531-8585

19 April 1976

To whom it may concern:

On Sunday, April 4, 1976 Reporter Pamela Flynn and Cameraman John Stromquist witnessed the following persons make a dive into Lake Waiau:

Tom Adams
Geoff Daigle
Glenn Egstrom
Ron Nolan
Mick Mathewson
Mark Murray
Dave Norquist

Lake Waiau is located near the summit of Mauna Kea at an altitude of 3,970 meters. It is purported to be the highest lake in the world.

Pamela J. Flynn

Pamela J. Flynn

John Stromquist

John Stromquist

CHAPTER 12
Whale Shark Rescue

Sea Frontiers

A copy follows of the original story published in
Volume 24 Number 3 May-June 1978

Mini, the Friendly Whale Shark

By Ron S. Nolan *and* Leighton R. Taylor

One of the biggest "fish stories" since Jonah and the whale is an account of Mini, a whale shark *(Rhincodon typus).* Many seafarers have had the pleasure of encountering this species, but only at unpredicted times and for brief periods. Usually the whale shark is found in open water and either sounds or drifts away from curious humans. Mini, however, was landlocked in a shallow lagoon, permitting the authors weeks of intensive observations.

The whale shark is aptly named because of its immense size and whale-like shape. It is the largest of living fishes and is so recognized by the *Guinness Book of World Records* (1975) which states that one individual measuring 59 feet long and weighing about 90,000 pounds was caught in a bamboo fish trap at Koh Chik, in the Gulf of Siam, in 1919. It is restricted to the tropical waters of the Atlantic, Pacific, and Indian oceans, and sightings of large sharks in cool temperate waters are probably of basking sharks *(Cetorhinus maximus).*

59 feet = 18 meters; 90,000 pounds = 40,823 kilograms

In spite of human interest in whale sharks, little is known about these remarkable creatures. Only rarely are they seen and even less seldom when a scientist and his equipment are at hand. Not even their mode of propagation is known, for the data that exist seem contradictory. For example, a large football-sized egg case containing an embryo whale shark was dredged from 31 fathoms in the Gulf of Mexico. Since the yolk sac had nearly been consumed, the examining biologist speculated that the embryo was near hatching stage, and it was concluded that the species was oviparous, depositing its eggs on the sea floor. More recently, however, a free-swimming juvenile whale shark, still bearing an umbilical scar, was captured in a tuna purse seine in the Pacific. This raises the possibility that whale sharks may be livebearers like many other sharks, and that the Texas egg and embryo may have been prematurely discharged by the mother.

The feeding habits of whale sharks also remain an enigma. The examined stomachs of several specimens contained large amounts of fresh algae, yet the scientific literature indicates that whale sharks feed by straining plankton in the manner of the baleen whales. Dr. C. McCann in a paper published in 1954 gave evidence that whale sharks are vegetarians, relying upon floating algae for sustenance. Whale sharks usually are seen slowly cruising at the surface, and McCann thinks that the warm surface temperatures may assist the shark's digestive processes. To add to the confusion, whale sharks have also been observed by scientists to ingest small fishes. Could these animals be more omnivorous than previously believed?

Rare Opportunity

With so much still to be learned about whale sharks, the authors welcomed the opportunity to observe and photograph a whale shark that was trapped in the lagoon at Canton Island, a small coral atoll in the Phoenix Islands group about 200 miles south of the equator. Canton has been under joint American-British control since 1937 and, in the last few years, has served as a missile tracking station for the United States Air Force, with Britain's approval. With the cooperation of the Air Force, a joint expedition, consisting of representatives of the Waikiki Aquarium in Hawaii and a private sea research firm called ORCA, set out to study the animal and attempt to free it if possible.

The expedition was arranged by Lt. Col. Philip Stack, the military commander of Canton, and a marine biologist. Stack explained that the shark swam into the entrance of the lagoon during a storm, and that, during the high tides that accompanied the storm, the animal had become lost in an intricate maze of coral channels. For three weeks the shark had routinely navigated a very restricted pond, separated from the 40-square-mile lagoon by a narrow channel. By the time the scientific expedition arrived, Canton residents had become concerned about the shark's well-being because it still refused to gain its freedom by passing through the channel leading to the lagoon.

Within an hour of landing at Canton, members of the expedition waded over the shallow reef flat to the spot where the shark was trapped. From a

200 miles = 322 kilometers

40 square miles = 104 square kilometers

All photographs by Ron S. Nolan

distance the enormous, sinister looking, black tail fin was spotted, as the shark slowly swam around its confinement. The scientists cautiously entered the murky water of the pond and dove in front of the slowly approaching shark. At first they could see nothing but opaque green water, but suddenly a dark submarine-sized object moved effortlessly beneath them. The sheer bulk of the shark was staggering; the head was estimated to be at least 4 feet in breadth. On the first dive the shark was seemingly oblivious to the presence of the men, so on subsequent dives they summoned their courage and hitched brief rides by clinging to the fish's dorsal fin. This reinforced their impression of the shark's size because from that perch they could not even see the animal's head. The massive shark simply swam on, paying no attention to its hitchhikers.

The shark, a female, soon became known as Mini. Measuring Mini underwater called for considerable coordination between the lead diver who frantically stroked to keep even with the formidable snout while streaming a measuring line and the trailing diver who just as frantically dodged the

4 feet = 1.2 meters

MAMMOTH FINS *project out of the water overlying the 27-foot whale shark ironically dubbed "Mini." The prominent first dorsal fin, on the right, shows the characteristic spotted coloration. The upper lobe of the tail is on the left. The shark was trapped within the lagoon of the Pacific island of Canton.*

Mini asked for shrimp for lunch!

massive sweeps of the tail fin while trying to read the measurement. After a few exciting attempts, it was determined that Mini was 26½ feet long. As awesome as the animal appeared, it was only about half grown!

26½ feet = 8 meters

Canton's residents were concerned that Mini might be in danger of starvation, so a cook kindly donated 50 pounds of frozen prawns for an attempt to feed it. No one predicted, however, how enthusiastically the

50 pounds = 23 kilograms

SAVING OPERATIONS *are undertaken in an attempt to force Mini out of a small seawater pond and toward the larger lagoon and eventual freedom. A 300-foot barrier net is dragged to the pond (above), where it is stretched open behind Mini in preparation for the herding attempt (right). The whale shark resisted all human efforts of rescue, finally making its way out of the pond on its own.*

shark would respond. A small skiff was anchored directly over Mini's normal swimming circuit and each time the shark passed beneath it, handfuls of shrimp were thrown into its anticipated path. It immediately sensed the food and, on its third pass, it halted at the boat and surfaced. One of the men bravely jammed a handful of shrimp into its mouth. The shark, however, expelled the shrimp with a giant blast of water. Its ambiguous behavior presented a problem since Mini clearly was interested in the food, but rejected it. The shark continued to bump the skiff and on one memorable occasion nearly capsized it.

The next day the shrimp supply was restocked, and Mini repeated her feeding attempts with even more enthusiasm. It was amazing to see the huge fish approach the boat and stick several feet of its head out of the water with its mouth wide open! At one point an entire 5-gallon bucket of shrimp was poured directly into its mouth. Mini expelled the offering as before, but this time the shark immediately opened its mouth and ingested a large amount of water and suspended food. After, this, the man feeding Mini would wait until the shark had expelled the water from its mouth and then bucket shrimp into its waiting orifice. It was startling to see the animal behaving as a well-trained pet as it repeatedly thrust its

5 gallons = 19 liters

open mouth out of the water against the skiff. In less than a dozen attempts, this primitive vertebrate had become trained in a feeding ritual different from any encountered in nature. Eventually the men thought it might be possible to lure Mini out of the pond with food, but every time the shark approached the channel it turned away. As a last resort, forceful means were used to try to save it.

Would-Be Rescue

A barrier net, 300 feet long, was assembled from smaller cargo nets and was stretched full-length across a narrow section of the pond trapping Mini in the channel end. Prior to the rescue attempt, Mini was marked with an aluminum harpoon tag at the base of the dorsal fin in an effort to gain information on whale shark movement. The tag stated, "Please record length of shark and locale of capture and send to Waikiki Aquarium."

On officially declared "Free Mini Day" all hands turned out to help draw the net towards the channel. The plan was to slowly restrict Mini's free-swimming space until the shark had no choice but to try the previously shunned channel. Excitement rose amidst an air of confusion (only half of the Samoan assistants spoke English) as the net drew nearer the escape channel. The shark was finally constrained in an area only slightly larger than its body. It sank from sight but still refused the channel. Suddenly, Mini charged the net dragging many of the determined saviors with it. At the end of four days of effort and plannings by the would-be rescuers, Mini was once again nonchalantly navigating its familiar route in the pond. The netting rescue attempt was stopped as the men took a much needed rest, but when observers returned to check on the shark they found it missing. It had finally tested the channel, without human help. A few days later Mini was sighted in the lagoon, peacefully meandering through coral-reef channels.

300 feet = 91 meters

Some members of the expedition to save Mini returned to Canton two months after the first encounter with the shark and soon found it in the far southern reaches of the lagoon, 10 miles from the pond where it had been trapped. As soon as the shark spotted the skiff, it swam over and nudged it, just as it had done in the pond.

Look for the Tag

Today, Canton is renowned throughout the Pacific for its pet whale shark. As of January, 1977, it was estimated that Mini had been in the lagoon for 14 months. The animal had been sighted in all sections of the lagoon and presumably had encountered the major reef passes, but it appeared to be in good condition and evidently was finding sufficient food resources within the confines of the rich lagoon. By now, Mini may have left the hospitality of Canton or may have been involuntarily swept out by strong seaward tidal currents. So, if IOF members who are cruising the Pacific sight a whale shark with a shiny tag protruding from its back, they are requested to report the location, date, and other observations to the authors at the Waikiki Aquarium, 2777 Kalakaua Avenue, Honolulu, Hawaii 96815. □

10 miles = 16 kilometers

CHAPTER 13
Ocean Thermal Energy Conversion (OTEC)

In 1979, I received contracts from the National Marine Fisheries Service and the State of Hawaii to develop methods of monitoring the effectiveness of installing aggregating buoys off the Kona Coast in order to attract pelagic game fish—like tuna and mahi-mahi. At that time there were no cameras available that offered underwater time lapse coverage, so I developed my own system which is pictured above.

The camera was also used by the world's first Ocean Thermal Energy Conversion (Mini OTEC) which employed over 2,000 feet of piping to transport cold water from the depths to the surface and then use the temperature differential to turn turbines and generate electricity beginning on August 2nd, 1979.

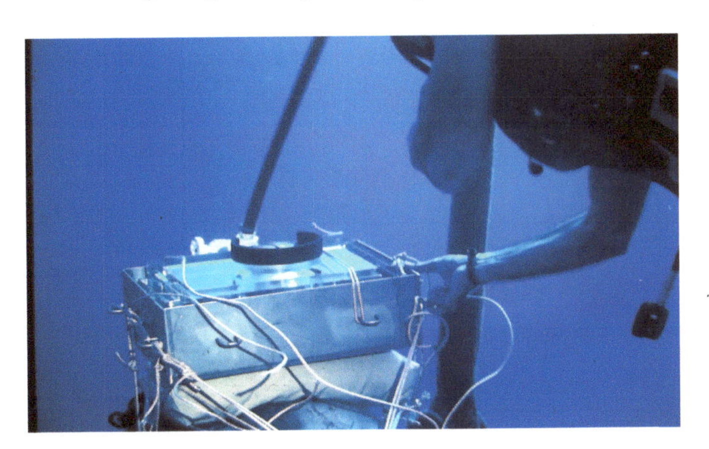

In the images above, the camera is being attached to the top of the pipe that extended to the depths below. It was programmed with instructions to take a picture at preset time intervals on 35 mm negatives that required special processing like movie film.

I was honored to participate in the world's first MINI-OTEC venture which produced power beginning on August 2nd, 1979. Plant operations concluded on November 15th, 1979 after successfully demonstrating the world's very first net positive, closed cycle at sea OTEC operation. Congratulations to the engineers involved and the State of Hawaii for providing initial funding for the project ax well as my participation.

OTEC-1 Pipe
Ready to Launch

My next consulting agreement was to provide services for the second generation of OTEC (OTEC–1) which was the larger scale version of the MINI-OTEC. My job was to provide photographic documentation of the growth of marine organisms on the hull of the mother ship.

The OTEC-1 Home Base
RV Ocean Energy Converter

Upwelling and discharging nutrient-rich water from the ocean depths into the surface waters stimulated the growth of sponges and tube worms on the mother ship's hull.

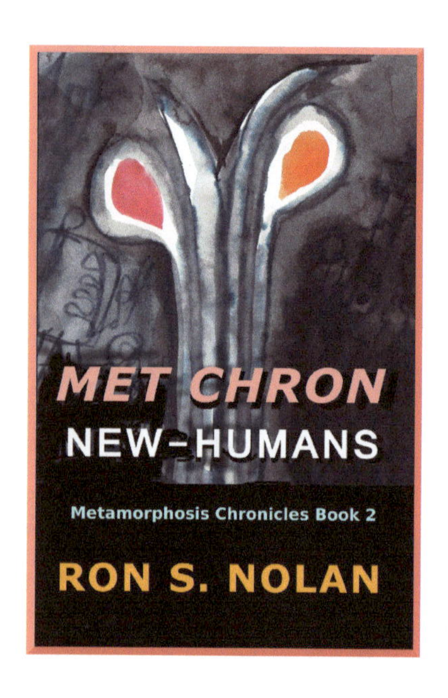

Readers of my MET CHRON series of science fiction novels may recall that Quant's plan is to use energy from their newly constructed OTEC plant on Majuro Atoll in the Marshall Islands. The idea is to generate electricity to power the Sky-Hook which transports warm seawater into space where it freezes into solid ice and then is returned to Earth's over heated seas and help ease the rapid rise in sea levels.

CHAPTER 14
Journey to the Desert

During my second year of graduate studies at Scripps, and somehow dealing with six months of Army National Guard tank gunner training during the below zero winter at Fort Knox Kentucky followed by two weeks stuck in a M48A2C tank blasting off ear throbbing 90 mm rounds in the raging heat of August in the heart of Death Valley. the tank shell was so hot you could fry eggs on the surface, it was prime time for a vacation break!

The trip through the desert of Baja Mexico took place 50 years ago and in those days all I had was a map and a wad of pesos that I had traded for U.S. Dollars at the border. At that time, the paved road down the coast (MEX 1) had not yet been built; this added to the adventure. To me it was all about the photo opportunity and my new Pentax camera that I had purchased a few months earlier on the Antipode Expedition from Osaka to Manila.

So here we go, extra gas cans strapped to the sides. Spare tires. Super bright lights above the windshield...imagine a trip today when you don't have a cell phone. At least, I knew some Spanish since my father William J. Nolan was a professor of Latin American Affairs and I had spent a year as an exchange student at Colegio Bolivar in Cali, Colombia. This part of the book will mainly present photographs taken on the trip with a little dialog, so enjoy!

Now, fifty years later, the trip seems like it took place on a Hollywood movie set. It was totally awesome with hardly anyone around except in the small villages.

On the road again...

El Rosario

We purchased gas and food from the rancheros on the way; kids laughing, running alongside and waving hello as we passed by.

Approaching La Paz

One Week Later...Back to School

CHAPTER 15
Journey to the Sea

I had the opportunity to spend several days on the Cynara when it stopped by Enewetak on its way to its new home in Japan. Built in 1927. One hundred feet long. She flies like a bird with the wind.

Long Live the Cynara!

Ron McConnaughey performed an amazing job of navigating the high seas on our voyage to Guadalupe Island, with the famous Scripps ichthyologist, Dr Carl Hubbs. Ron used the Boston Whaler to transport Bob Warner and I to the island where we spent a week diving and exploring. Oh how great that trip was!

The Whaler usually was on the Scripps pier and lowered down or raised with a hoist so we could dive in the submarine canyon right next to the Scripps campus.

When Ron found out that we would like to survey the marine life around Guadalupe Island, he asked that the Whaler be brought along because there would be no launching ramps or beaches, only sheer, lava rock faces.

The solution he used was to quickly land the Whaler onto the shore with an incoming wave and Bob and I would grab gear just as Ron used a set of oars to row the boat off before it became trapped by the rocks as the water drained back into the ocean.

This was an amazing display of courage and talent that I had never seen before and that had to be repeated many times at the outset and when the ship with Ron and the Whaler on board returned after collecting deep sea fishes from the local submarine canyon. I liked the Whaler so much that I bought one.

Dan Cheney, Ph.D. Marine Biologist from the University of Hawaii, Hilo Campus is at the helm of my newly purchased Boston Whaler which was in immaculate shape because its prior service was commercial fishing at South Point. The dog lovingly named "Rat" was an Australian "red healer" who just showed up one day at my house in Puako. He was extremely intelligent and more than willing to accompany me on my various ventures.

The lava island was in our backyard.

Boat Heading into the Mystic

CHAPTER 16
Beauty In Flowers

Wildflowers on the California Coast

Vibrancy in Color

Sex

Love

CHAPTER 17
Beauty in the Sea

Giant groper claims wrecked ship as its territory.

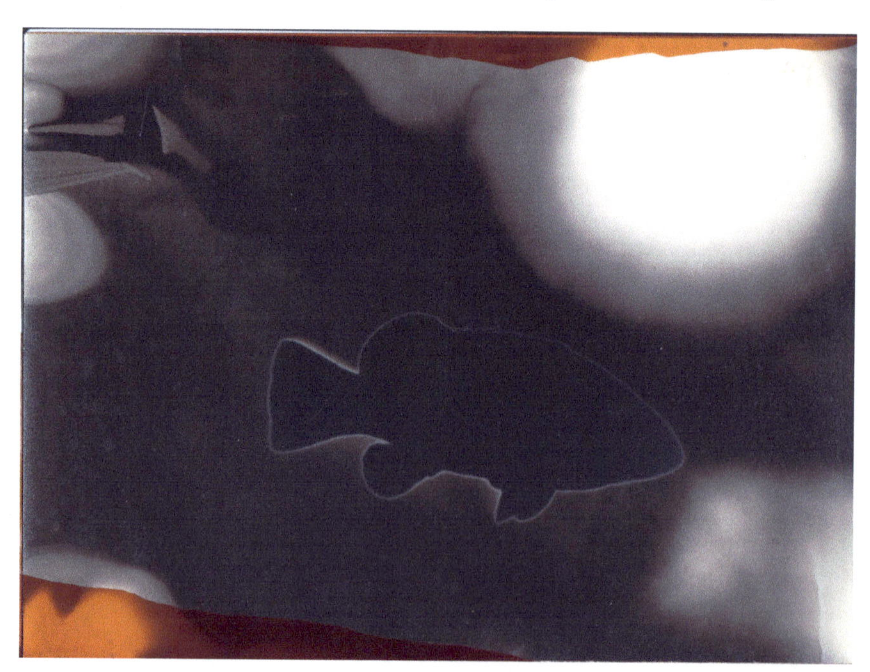

School of Chromis hiding in a coral colony.

Beauty is actually a warning signal in venomous lionfish.

Manta ray feeding

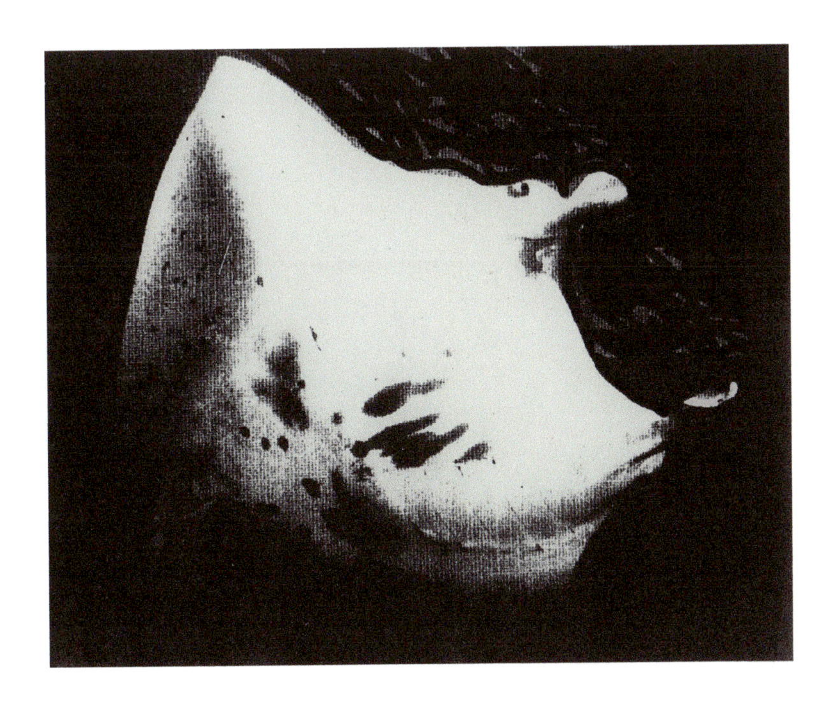

A master of camouflage!

CHAPTER 18
Beauty in Forests

CHAPTER 19
Beauty in Flowing Waters

Clearwater River, Idaho

Big Wood River, Ketchum, Idaho

CHAPTER 20
A Few of My Favorite Pictures

Northern California Coastline & Sea Pine

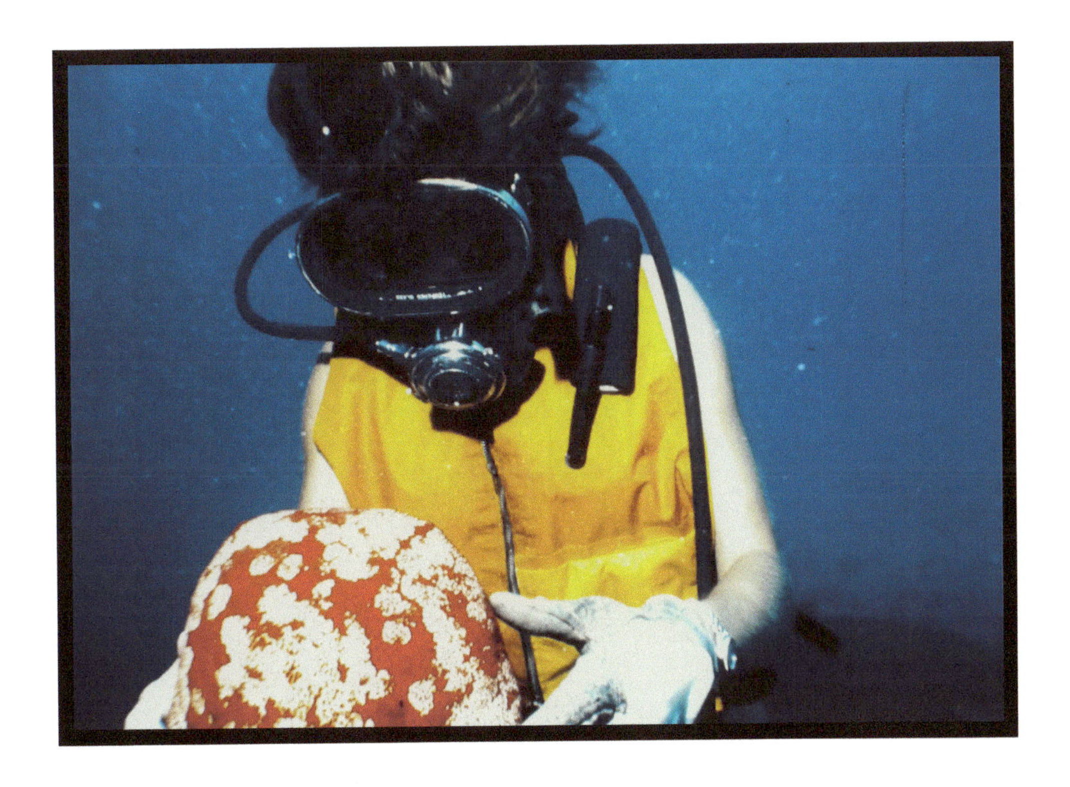

Unicornfish

Skiing on the Big Island's Mauna Kea

This house is a very, very fine house with...

The Nature of Spirituality

Wrasse & Butterflyfish

Near surface dwelling coral in which polyps provide homes for photosynthetic zooxanthellae which in turn provide food to the coral in exchange for a safe place to live. The relationship is an examle of "mutualism."

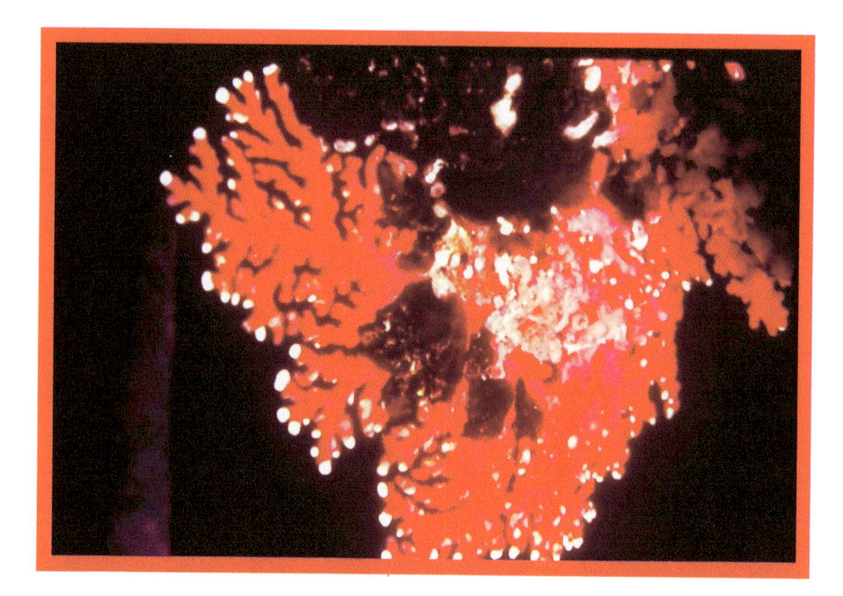

Corals like this live beneath the photic zone and are carnivores.

Taken at a depth of 265 feet.

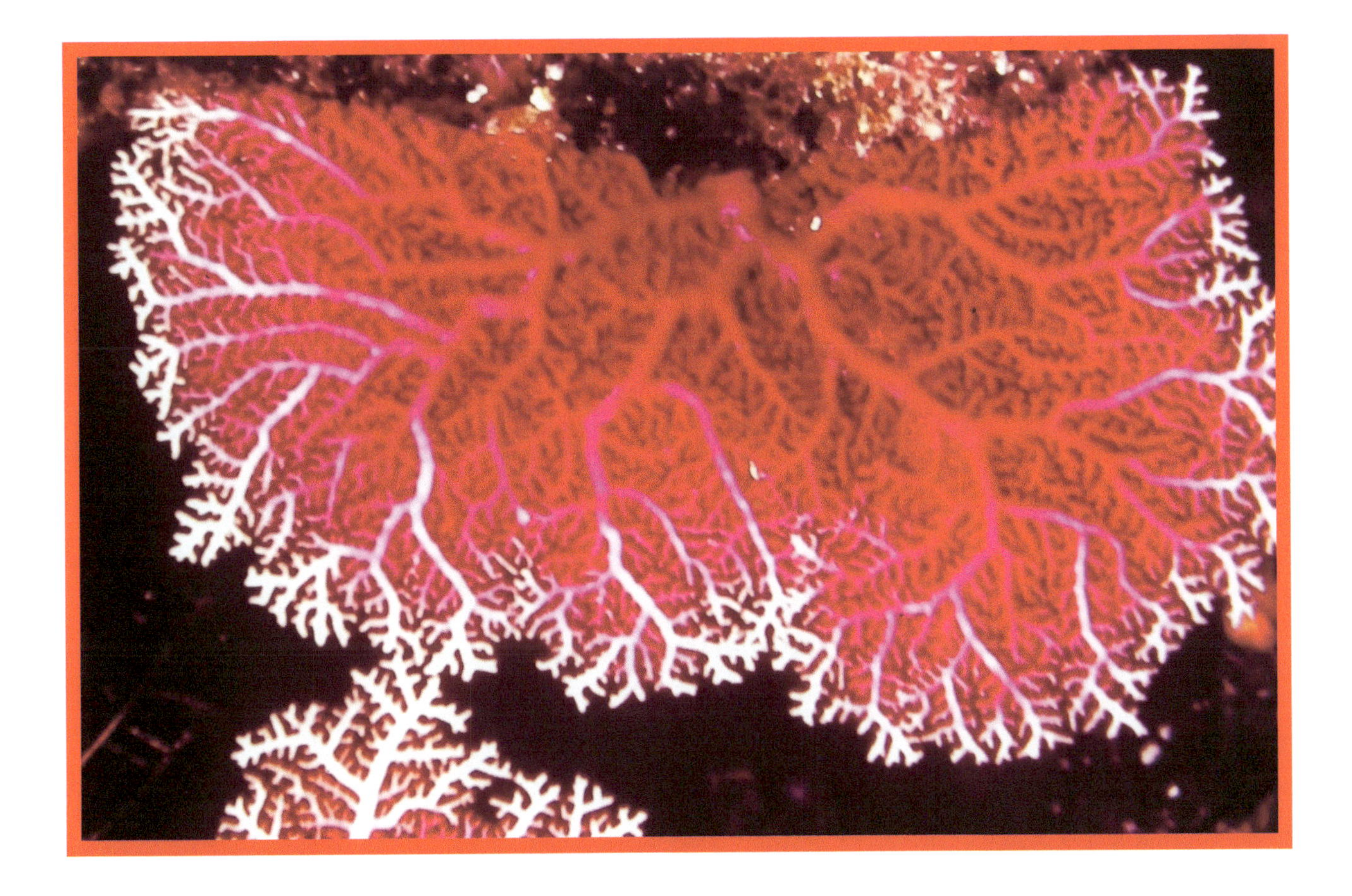

Waianae Coast
Oahu

Healthy Coral Colony

Crown of Thorns Feasting on Coral

Real Beauty in Nature

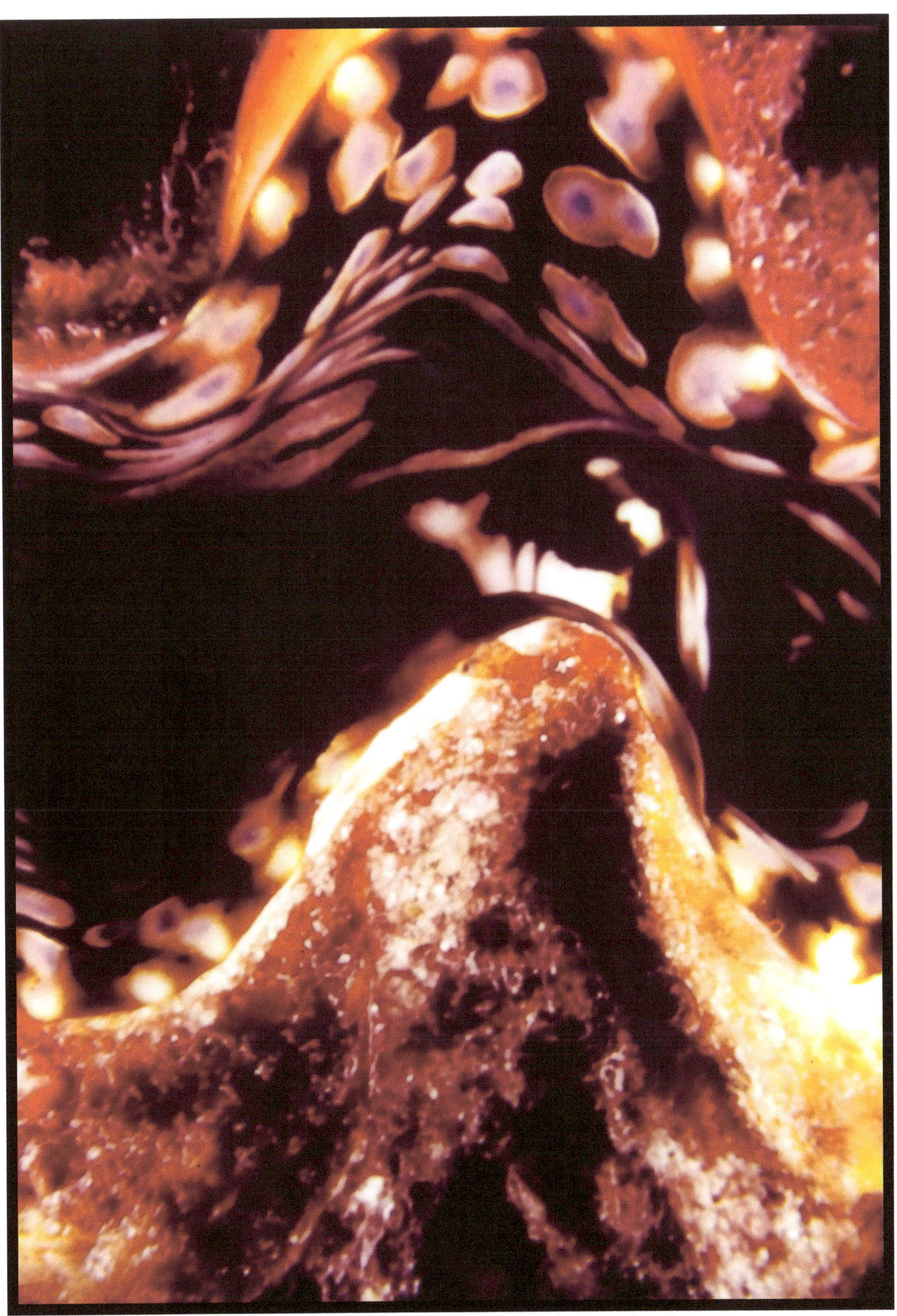

Giant clams also harbor zooxanthellae.

Looking Forward

In part two of my next book of recollections, the journeys continue with expeditions to interesting and unexpected places. For example, exploring ship wrecks in Truk Lagoon and starting a shrimp farm on Molokai.

In the mid-1990's, I became fascinated with the potential for employing computers in media delivery and became a Research Associate at both UCSC and Apple which gave me access to the most advanced hardware and software at that time. Dr. Patrick Mantey and I used these advanced to produce one of the first CD-ROM programs called "*Computer Music: an Interactive Documentary.*"

My involvement with UC Santa Cruz had an amazing twist which I will elaborate on in Book 2 and how I came to know the inventors of the laser and the semiconductor. I hope you enjoy reading this book as much as I have in writing it. I am fortunate to have had such a wild ride.

To join my mailing list, send a message to nolan@planetropolis.com**.**

ABOUT THE AUTHOR

Ron S. Nolan, Ph.D. lives in Aptos, California near the sunken ship at the end of the pier in SeaCliff Beach. He spends his days working out, running, writing and performing tech patent research–quite a leap from his early days in Western Kansas where he shared the farm outhouse with a nest of half frozen rattlesnakes and learned to read by the light of a Coleman lantern! To learn more about his latest novels and screenplays, please visit…

Planetropolis Publishing
www.planetropolis.com

MET-CHRON SANCTUARY

Metamorphosis Chronicles Book 1

Astra, a head-turning, Brazilian girl in her mid-twenties is not only beautiful but also brilliant...and in big trouble! After discovering a key to the genetic aging clock that could dramatically increase the human lifespan, she is tracked down by a psychic who delivers a stern warning that she must work to heal the planet before adding to the over population crisis by allowing a select few to live longer lives.

In the year 2029, the terminal impacts of global warming are having disastrous effects upon ecosystems and global tensions have escalated as countries fight to extract the last barrel of oil. Astra's new mission is to assemble cryogenic repositories ('Arks') of frozen plant and animal embryos to preserve them for the future. However, she is opposed by a fanatical religious group that will do anything to stop her. But is it the Ark that they really want...or something hidden within? Either way, they will have to go to a mining base on the moon to find out.

Available in eBook, Softcover & Hardback
www.planetropolis.com

MET CHRON NEW-HUMANS

Metamorphosis Chronicles Book 2

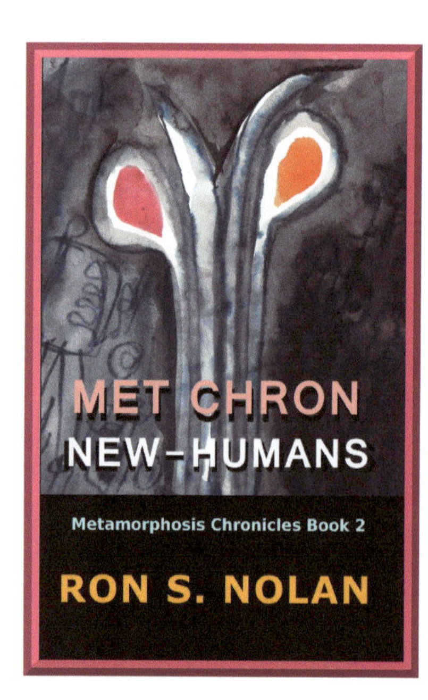

Set in the year 2030, a Sci-Fi technothriller of a world near the global warming tipping point and humanity's survival is threatened, from the author of *Met-Chron Sanctuary and Telepathic Dolphin Experiment.*

Timing is everything. If the intern at the SpeeZees Lab on the Moonbase doesn't accidentally select the wrong code sequence in a training program, the world's first New-Human will not be 3D printed and brought to life....and without Chron's genius, the strategy of developing a space elevator to convey pods of seawater into space and generate snowfall over tropical seas, coastal cities around the world will face catastrophic flooding. Furthermore, if Chron does not assist her longevity research, Dr. Astra Sturtevant will not make her game clinging discovery of how to control the genetic clock and arrest the aging process. Meanwhile, the AI androids that had mutated in the high RAD zones following the nuclear detonations in the Bay Area, have launched campaigns to gain their independence and plan to annihilate all organic humans. The innovative pioneers that operate the Deep Space Mining Moonbase have a plan that will solve humanity's problems. The question is will anyone listen?

TELEPATHIC DOLPHIN EXPERIMENT

During a life-long search to scientifically document paranormal phenomena, Dr. Sandra Grant discovers that dolphins offer ideal research subjects. Through her persistence and the aid of a government contractor, a pair of twin dolphins is made available to her by a paranoid general intent upon the ultimate destruction of the USSR, but facing a public relations nightmare due to his support of a recently exposed secret military program to use marine mammals as weapons.

General Pratt Houston's intelligence sources indicate that Russia has an overwhelming arsenal of nuclear weapons. Driven mad by the government's past reluctance to use full military force in Viet Nam, and with the refusal of Congress to use nukes against the Iraqis, General Houston forces a computer programmer to create a virus designed to wreak havoc on the Soviet Defense Network. Designated as ANX, the virus will penetrate the defense network and disrupt their communication systems, When Russia panics and launches their missiles out of desperation, ANX will corrupt their guidance systems and cause them to misfire. The subsequent U.S. counterattack will permanently solve the arms imbalance--at least according to the General's twisted thinking.

However, through a bizarre chain of events, the ultimate fate of humanity depends upon the determination and resourcefulness of Dr. Grant and her telepathic dolphins to thwart the General's sinister plan.

Available in eBook, Softcover & Hardback
www.planetropolis.com

FREEDOM RIDE

From the wheat fields of Kansas to surfing the immense waves on the shores of Planet Nuptia, the author takes readers on a journey which may be referred to as the "future meets the past."

In "Earth Boy" a child is held captive on Europa and in the "Longevity Gene" a geneticist searches for a gene complex that may hold the key to immortality while stock car drivers race around a track in Alabama and discover something totally unexpected.

As the book title suggests, searching for and struggling to acquire independence is the underlying theme of Freedom Ride's short stories which concludes with poems about seagulls, crows, ship wrecks, dangerous sharks and young love that makes today seem like yesterday.

ALIEN PORTAL
Metamorphosis Chronicles
Book 3

Send me an e-mail at nolan@planetropolis.com and I will keep you posted as new titles become available.

ALOHA from Ron S. Nolan

CPSIA information can be obtained
at www.ICGtesting.com
Printed in the USA
LVHW072136181122
733442LV00002B/5

* 9 7 8 1 7 3 7 9 6 8 1 5 3 *